ALPHA BOOKS

HURRICANES AND STORMS

NICOLA BARBER

Evans

EVANS BROTHERS LIMITED

This book is based on **Repairing the Damage** HURRICANES AND STORMS by Clint Twist, first published by Evans Brothers Limited in 1992, but the original text has been simplified.

Evans Brothers Limited
2A Portman Mansions
Chiltern Street
London W1M 1LE

© Evans Brothers Limited 1993

All rights reserved. No part of this publication may be reproduced, stored in a retrieval system or transmitted in any form or by any means, electronic, mechanical, photocopying, recording or otherwise, without prior permission of Evans Brothers Limited.

First published 1993
Reprinted 1996

Typeset by Fleetlines Typesetters, Southend-on-Sea
Printed in Spain by GRAFO, S.A. – Bilbao

ISBN 0 237 51325 0

Acknowledgements

Editor: Su Swallow
Language adviser: Suzanne Tiburtius (East Kent Integrated Support Services)
Design: Neil Sayer
Production: Jenny Mulvanny

Maps: Hardlines, Charlbury
Illustrations: Andrew Calvert

For permission to reproduce copyright material the author and publishers gratefully acknowledge the following:

Cover (top) Erwin and Peggy Bauer, Bruce Coleman Limited (bottom) Keith Kent, Science Photo Library
Title page (Lightning, Arizona) Keith Kent, Science Photo Library **p4** Omikron, Science Photo Library **p5** W Broadhurst, FLPA **p6** Steve McCutcheon, FLPA **p7** D Nicholls/ECOSCENE **p8** (top) Hasler and Pierce, NASA GSFC/Science Photo Library, (inset) NOAA/Science Photo Library, (bottom) Frank W Lane, FLPA; **p10** (top) Howard Bluestein, Science Photo Library, (middle) R Steinan, FLPA, (bottom) M J Coe, Oxford Scientific Films; **p11** Erwin and Peggy Bauer, Bruce Coleman Limited **p12** Tom Van Sant/ GeoSphere Project, Santa Monica/Science Photo Library **p15** Keith Kent, Science Photo Library **p16** Fritz Pölking, FLPA **p17** Hoflinger, FLPA **p18** Barnaby's Picture Library **p19** Australia Information Service, FLPA **p20** Popperfoto **p21**(left) British Red Cross, (top right) Popperfoto/AFP, (bottom right) Gerald Cubitt, Bruce Coleman Limited **p22** James Stevenson, Science Photo Library **p23** Mary Evans Picture Library **p24** Dr Fred Espenak, Science Photo Library **p26** two eyewitness accounts taken from an article in The Times 16.9.88 by Alan Tomlinson, © Times Newspapers Ltd 1988 **p27** Bernard Bidault, GAMMA/Frank Spooner Pictures **p28** S Jonasson, FLPA **p29** David Parker, Science Photo Library **p31** (top) Cooper/ECOSCENE, (bottom) Ben Fawcett, Oxfam **p33** (top) D Hoadley, FLPA, (bottom) Harry Nor-Hansen, Science Photo Library **p35** Dr Jeremy Burgess, Science Photo Library (inset) Crown copyright **p36** Crown copyright **p37** European Space Agency/ESA/Science Photo Library, (inset) Dr P Menzel, Science Photo Library **p38** (left) Robert A Lubeck, Oxford Scientific Films, (right) National Centre for Atmospheric Research/Science Photo Library **p39** Vintage Magazine Company **p40/41** Norm Thomas, Science Photo Library **p42** Dr Bernard Stonehouse **p43** Mary Evans Picture Library.

The Crown copyright pictures on pages 35 and 36 are reproduced with the permission of the Controller of HMSO.

The author and publishers have been unable to trace the origin of the quotations on pages 17 and 20 in order to acknowledge them.

Contents

Introduction 4

All kinds of storm 6
Thunderstorms   Fire from the sky   Hurricanes
Giant storms   Whirlwinds   Sand and snow

How winds work 12
The earth's atmosphere   Heat and pressure   Prevailing winds
Thunder and lightning   Hurricane winds
The birth of a hurricane   Twisting air

After the storm 18
Clearing up   The worst storm

A night of storm 22
A storm begins

Gilbert's story 24
Diary of a storm   The beginning
The middle   The end

Protecting lives and land 28
Staying safe   Forecasting weather   Storm warnings
Hurricane parties   Protecting the coastline   Plastic seaweed
Sheltering from tornadoes

Weather forecasting 34
Reports from sea   Reports from the air   Satellites   Hailstones
Weather maps   Hurricane off course

Climate and weather 40
Warming the atmosphere   Controlling the weather

Glossary 44

Index 45

Introduction

Sometimes the air around us moves so fast that it blows away trees, houses and people. We call this moving air wind. When the wind blows hard and rain falls we call it a storm. There are many different kinds of storms. Some are small thunderstorms with lightning and thunder. Other storms are so huge that they cover half of an ocean.

▽ A bad storm hits Palm Beach, in Florida, USA.

TROPICAL SPINNING STORMS

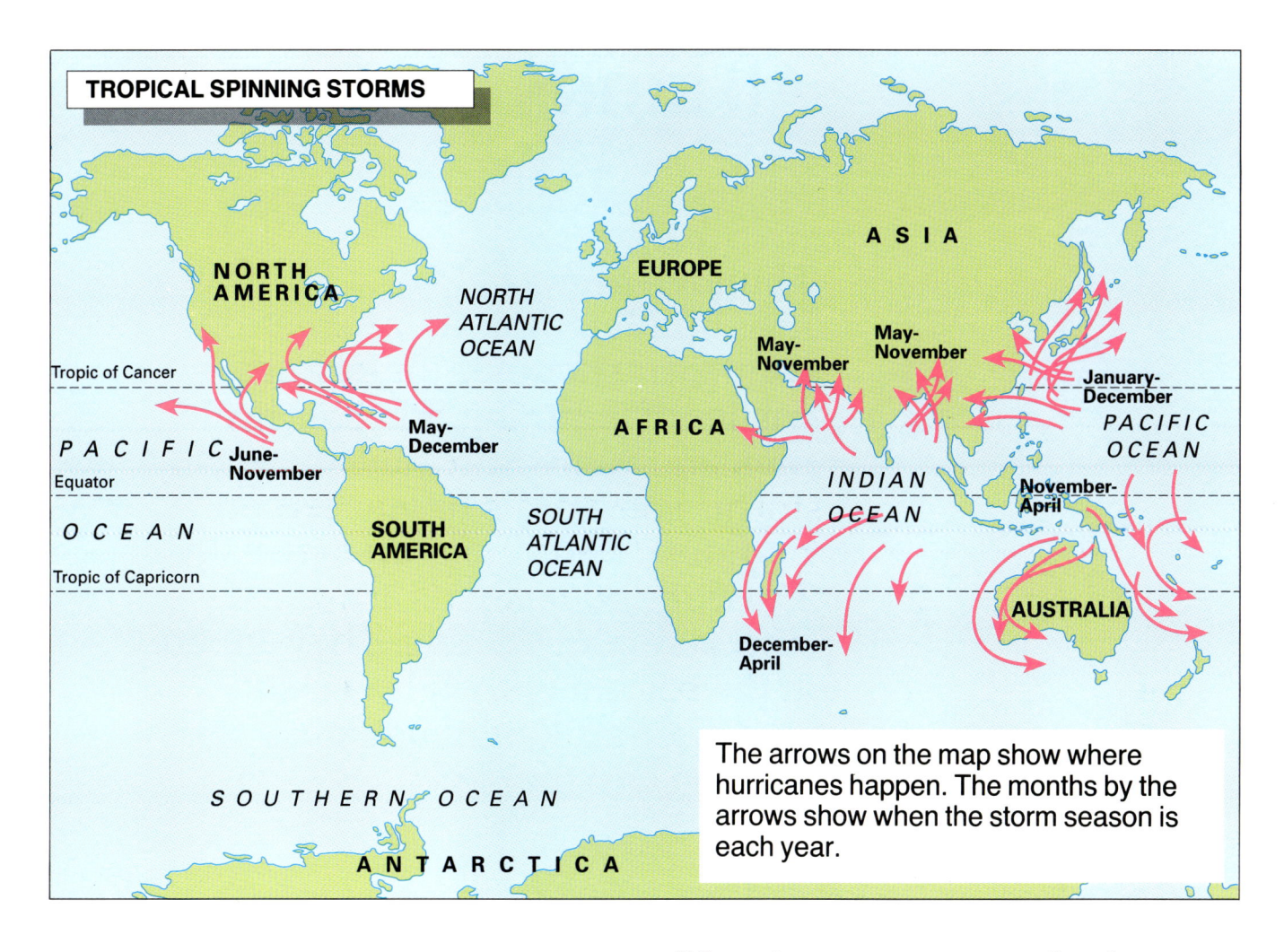

The arrows on the map show where hurricanes happen. The months by the arrows show when the storm season is each year.

▽ In a storm, strong winds and rain can flatten crops.

Hurricanes are very bad storms. Hurricanes can be thousands of times more powerful than thunderstorms.

Most places on earth have storms. There are storms even in the deserts, and at the North and South poles. Some places do not have bad storms very often. But in other places there is a storm season every year.

All kinds of storm

Thunderstorms
In a thunderstorm there are strong winds, crashes of thunder and flashes of lightning. There is heavy rain, and sometimes hailstones fall.

Thunderstorms happen when thunder clouds build up in the sky. These thunder clouds are called cumulonimbus clouds. They are grey at the bottom with lighter cloud above. Cumulonimbus clouds reach high up into the sky.

▽ A cumulonimbus cloud shows that a thunderstorm is on the way.

◁ Fire helps the Banksia bush to grow. These pictures show the flower (left), the seed pods (top right) and a new plant (above).

Sometimes hailstones form inside a thunder cloud. The biggest hailstones can damage cars and hurt people. Luckily, hailstones are usually smaller than this.

Fire from the sky
When lightning hits the ground it can start a fire. Every year, many fires are started by lightning. Fires destroy large areas of forest. But fires also help some plants to grow. The Banksia bush grows in Australia. The flowers of the Banksia bush turn into pods with seeds inside. After a fire, the seeds fall out of the burnt pods and grow into new plants.

Hurricanes

A hurricane is a much bigger storm than a thunderstorm. A hurricane is a tropical spinning storm. It can measure up to 300 kilometres across. In a hurricane the wind blows round and round very fast. But in the middle of the hurricane there is a place where the wind is gentle. This place is called the eye of the hurricane. The wind also blows heavy rain along with the hurricane.

▷ An instrument called an anemometer measures the speed of the wind.

△ (Inset) A picture taken from a weather satellite shows the spinning cloud of a hurricane. Another picture (above) shows the eye of the hurricane.

A hurricane is the name given to a tropical spinning storm in the Atlantic Ocean. In the Pacific Ocean these spinning storms are called typhoons. In the Indian Ocean they are called cyclones. Every year there are about 90 of these storms. Some happen out at sea where they cannot do any harm. But some reach land where they can kill many people.

Giant storms

Hurricanes are not the most powerful storms on earth. Giant storms happen in the middle of the Atlantic and Pacific oceans. These storms can be more than 3000 kilometres across. These giant storms also spin, but much more slowly than hurricanes.

The Beaufort Scale
The speed of wind is often measured on the Beaufort Scale. On this scale, a wind blowing at four kilometres per hour is called Force 1. A wind blowing at ten kilometres per hour is called Force 2, and so on.

Force	Approx speed (kph)	Effects
1	4	smoke drifts
2	10	leaves rustle
3	17	twigs move
4	26	small branches move
5	36	small trees sway
6	48	large branches move
7	58	whole trees move
8	72	small branches break
9	85	houses damaged slightly
10	98	trees broken
11	112	widespread damage
12	117+	violence and destruction

Whirlwinds

The fastest winds on earth are in tornadoes. A tornado is a thin tube of spinning air reaching from the sky down to the ground. The winds inside a tornado can spin at up to 500 kilometres per hour. In the United States, a tornado once lifted a railway train off its tracks. The tornado carried the train 50 metres before dropping it to the ground again.

Tornadoes happen all over the world although they are not usual in India and Africa. Tornadoes in Europe are mostly quite small.

△ A spinning tornado picks up soil and dust from the ground.

△ Houses and trees are damaged when a tornado passes by.

Tornadoes in the United States can be much bigger, and can do a lot of damage. In the south of the United States there is an area called Tornado Alley. It stretches from Texas to Missouri. There are more than 300 tornadoes every year in Tornado Alley.

Tornadoes usually last for about 15 minutes, although some go on for hours. The spinning air in the tornado looks white until it reaches the ground. Then it turns brown as it sucks up soil and dust from the ground.

Sand and snow

Thunderstorms, hurricanes and tornadoes all have very strong winds. But sometimes gentle winds can be as dangerous as strong winds, especially when the winds carry sand and snow.

When the wind blows sand it is called a sandstorm, or a dust storm in the United States.

Sandstorms happen in

▽ A sandstorm in the Indian Desert

deserts. There are two kinds of sandstorm. A large sandstorm is called a khamsin. A khamsin happens when a strong wind blows across dry sand or dust. The wind blows the sand or dust into the air. A khamsin can last for days, covering roads and fields with sand and dust.

The other kind of sandstorm is called a haboob. A haboob lasts for only a few hours. It usually happens just before a thunderstorm. The sand and dust are blown up very high by the winds from the thunderstorm.

When the wind blows snow it is called a blizzard. The wind piles the snow into layers, called snowdrifts.

▽ A blizzard blows in Alberta, Canada

How winds work

The earth's atmosphere
All around the earth is a layer of air called the atmosphere. The air in the atmosphere is made up of a mixture of gases. High up in the atmosphere

▽ This view of the earth is made out of thousands of pictures taken from a satellite.

the air is very thin. This is because the gas molecules are far apart. Close to the earth the air is thicker. The gas molecules are packed tightly together. The thickness of the air pushing down on the surface of the earth is called atmospheric pressure.

Heat and pressure
Light from the sun heats the atmosphere around the earth. But not all parts of the atmosphere get the same amount of heat. The air above the equator gets sunlight all year. But the air above the North and South poles gets sunlight for only part of the year.

When sunlight heats the atmosphere the gas molecules in the air move further apart. This makes the air thinner and the atmospheric pressure gets less. In places where the air is heated there is low pressure. There is a band of low pressure around the equator. In places where the air is colder, there is high pressure. There is high pressure at the North and South poles. Air moves from the high pressure areas to the low pressure areas. We call this movement of air wind.

▽ The layers of the earth's atmosphere

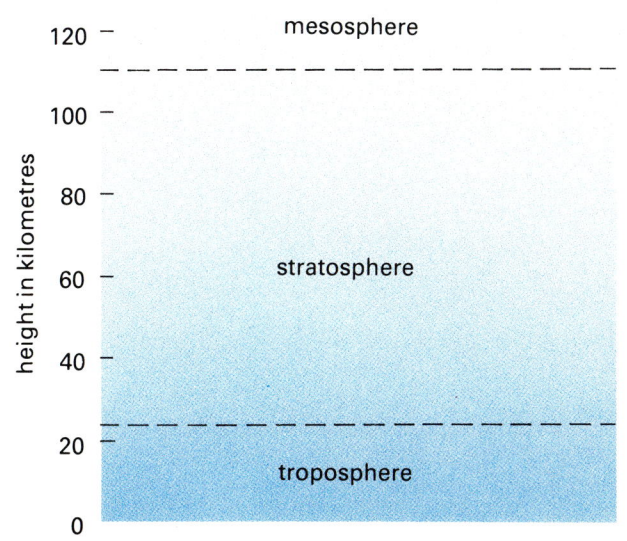

North Pole

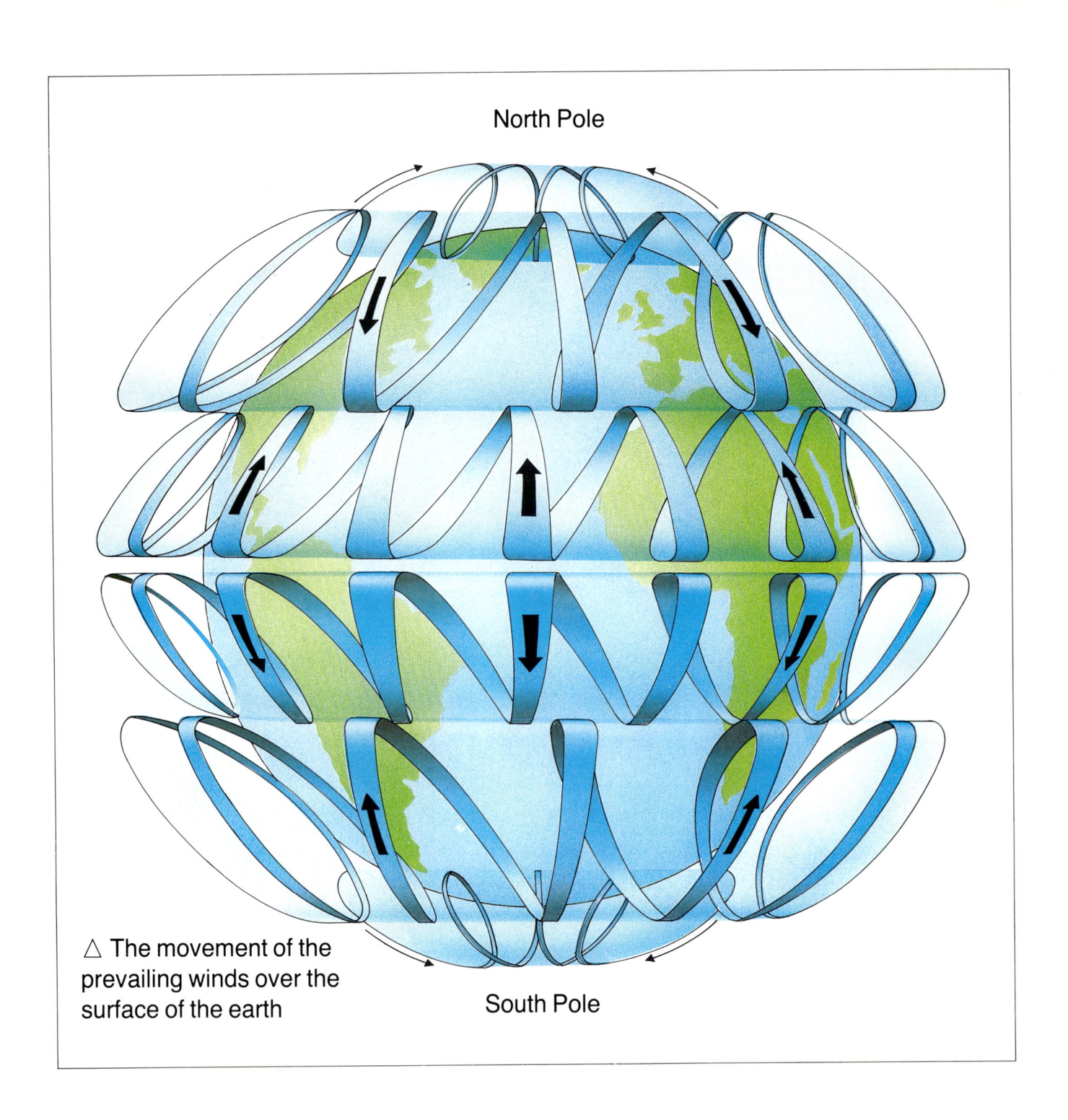

South Pole

△ The movement of the prevailing winds over the surface of the earth

Prevailing winds

Winds blow across the surface of the earth from areas of high pressure to low pressure. But the earth is spinning round very fast. This makes the winds move on a curved path. The pattern made by the winds is shown in the picture above. These winds are called prevailing winds.

Thunder and lightning

A thunderstorm often starts when cool air meets warm air. The cool air stays close to the ground. It pushes the warm air upwards. The warm air forms a cumulonimbus cloud (see page 6). Inside the cloud, electricity builds up. The electricity escapes as a bright flash of lightning. It moves between the bottom of the cloud and the ground. The lightning heats the air as it moves. The noise of the air being heated is called thunder. Lightning and thunder happen at the same time, but you usually see the lightning before you hear the thunder. This is because light travels faster than sound.

▽ A thunderstorm in Arizona, USA

Hurricane winds
Hurricanes begin over a sea or an ocean. But they only form if the temperature of the surface of the water is higher than 26 degrees Celsius. This means that hurricanes form over tropical seas and oceans. Warm air rises from the surface of the water. Sometimes the air turns in a spiral, spinning faster and faster. This is the start of a hurricane. Hurricanes are carried by the prevailing winds (see page 14).

The birth of a hurricane
No one knows exactly how a hurricane starts. When a butterfly flaps its wings it could start warm air moving upwards. Or if someone clapped their hands they could disturb the air. It is impossible to tell what small movements start a hurricane. This means that scientists cannot tell when and where a hurricane is going to start. It is also difficult to tell where a hurricane will go once it has started. Even with the fastest computers, scientists cannot say in what direction a hurricane will move for more than a few days ahead.

△ The movement of a butterfly's wings could be the start of a hurricane.

△ A waterspout off the coast of Italy

Twisting air
A tornado forms inside a thunder cloud. Sometimes a thunder cloud contains several tornadoes. When the tornado is spinning fast enough it touches the ground. It sucks up soil and dust, and then it dies away. If a tornado forms over water it is called a waterspout.

Tornadoes can destroy buildings and kill people. In areas of the USA where tornadoes happen often, houses have underground tornado shelters. People get into the shelters when they see a tornado coming. In 1928, an American farmer in Kansas stopped at the door to his shelter and looked back. This is what he described:

'Steadily the tornado came on, the end gradually rising above the ground. At last the great shaggy end of the funnel hung directly overhead. Everything was as still as death. There was a screaming, hissing sound coming directly from the end of the funnel. I looked up and to my astonishment I saw right up into the heart of the tornado. There was a circular opening in the centre of the funnel, about 50 or 100 feet (15–30 metres) in diameter, and extending straight upward for a distance of at least one half mile (800 metres) as best as I could judge. The walls of this opening were of rotating clouds and the whole was made brilliantly visible by constant flashes of lightning which zigzagged from side to side.'

After the storm

Clearing up

After a storm there is often a lot of damage. Leaves and small branches lie on the ground. Big branches are broken. Some trees may even fall over. Rivers are full of rainwater which washes soil away from the river banks. Houses are often damaged too. The slates on the roofs may blow off. Chimneys and walls sometimes collapse. Some wooden buildings blow away or fall down completely.

▽ Strong winds blew this tree over on to the van.

The first job after a bad storm is to help the people who have been affected by the storm. Fallen trees often block roads. The trees must be moved so that people can get to the areas hit by the storm. A badly damaged building could fall down and hurt someone. So the building must be made safe or knocked down. The power lines that carry electricity are often broken. These must be mended so that people have electricity in their houses again. Sometimes the water pipes are also broken and they must be mended too.

In cities, tall buildings are very strong so that they will not fall over in high winds. But storms can still do damage in a city. There was a bad storm in London in 1975. This is how one office worker described it:

'It started to hail. The hail turned to rain. There was some terrific thunder. After about an hour, it stopped raining. I walked down the hill. The street was full of water. By the time I got to my house, I was wading up to my knees. Fortunately, I lived on the first floor, my flat was dry. But all the people in the basement and ground floor flats were flooded out. Some friends had over two metres of water in their living room and had to be rescued by boat.'

Hurricane winds are much stronger than storm winds, so they do more damage. In 1974, a hurricane called Cyclone Tracy hit northern Australia. On Christmas Day it almost destroyed the city of Darwin. The hurricane winds damaged nearly all the buildings. The people of Darwin had to rebuild their city.

△ Cyclone Tracy destroyed nearly all the buildings in Darwin, northern Australia.

The worst storm

The worst storm in recent years happened in 1970. A cyclone started in the Indian Ocean. The spinning storm sucked up water from the ocean. Then the cyclone reached the mouth of the River Ganges. As the cyclone moved across the coast it dragged seawater with it. A huge wave of water hit the land. The water washed away people, animals and houses. More than one million people died. A farmer who survived the cyclone described what happened afterwards:

▽ A family escapes from the floods after another cyclone in 1974.

'About dawn the water began to go down. I could see bodies, hundreds of them, floating out to sea. At about nine in the morning the water finally went down. My farm looked like a desert. There was nothing left, but my family was all right.'

The area hit by the cyclone in 1970 is now called Bangladesh. Since that cyclone, the government of Bangladesh has built strong cyclone shelters. The shelters stand on thick, concrete stilts. This means that they are not washed away when a storm wave covers the land with

◁ A cyclone shelter (left) and mangrove forests (right). In 1991, planes flew over areas cut off by the floods to drop food to the people there (above).

water. In April 1991, another cyclone hit Bangladesh. This time, many people went to the cyclone shelters. They were able to survive the cyclone.

One problem is that cyclone shelters are expensive to build. Another way to protect the coast from cyclones is by planting mangrove forests.

Mangrove plants grow in marshy areas along the coast. They help to slow down the water in a storm wave. This means that the water does not flood as much land. Since 1970, the government has planted many mangrove trees along the coast of Bangladesh.

A night of storm

In 1987, the worst storm for 200 years hit the south of England. It happened on the night of the 15th–16th October. Nineteen people died in the storm. Most people were at home in bed when the storm struck. The roads were nearly empty. If the storm had happened during the day, more people could have died.

The storm did a lot of damage. Seven million people were left without electricity. The strong winds damaged many houses. The winds also blew about 15 million trees down. The trees blocked many roads and railway lines.

A storm begins
The storm of 1987 began in the Bay of Biscay, off the coast of France and Spain. On the other side of the Atlantic Ocean a hurricane was blowing across Florida in the USA. This hurricane was called Hurricane Floyd. A stream of air from Hurricane Floyd crossed the Atlantic Ocean. The air was moving very fast. When the air reached the Bay of Biscay it joined with the storm there. The air from the hurricane increased the

▽ The storm of 1987 damaged many buildings.

△ The great storm of 1703, off the coast of England

power of the storm in the Bay of Biscay. When the air in two storms meets like this, no one can tell what will happen. Sometimes the storm dies away. Sometimes it gets much more violent.

From the Bay of Biscay, the storm moved across the north of France. In Paris, a giant crane collapsed. The wind flattened crops and blew trees down. The highest wind speed of the storm was recorded on the coast of Brittany. The wind speed was 214 kilometres per hour.

The storm of 1987 was a terrible storm. But it was not the worst storm ever to strike England. The great storm in 1703 was even worse. It lasted a whole week and it killed 8000 people.

Gilbert's story

Diary of a storm

In the Caribbean Sea, storms are quite usual in late summer. But in September 1988, several storms joined together in the Caribbean. They formed Hurricane Gilbert. This was probably the most powerful hurricane for over 100 years.

Hurricane Gilbert lasted for ten days. It killed at least 400 people. It injured many thousands of people, destroyed houses, and flattened crops. This diary of Hurricane Gilbert tells how the storm moved across the Caribbean. You can look at the map on page 25 to see where all the places are.

▽ A picture of Hurricane Gilbert taken from a satellite on 14 September, 1988

HURRICANE GILBERT'S PATH

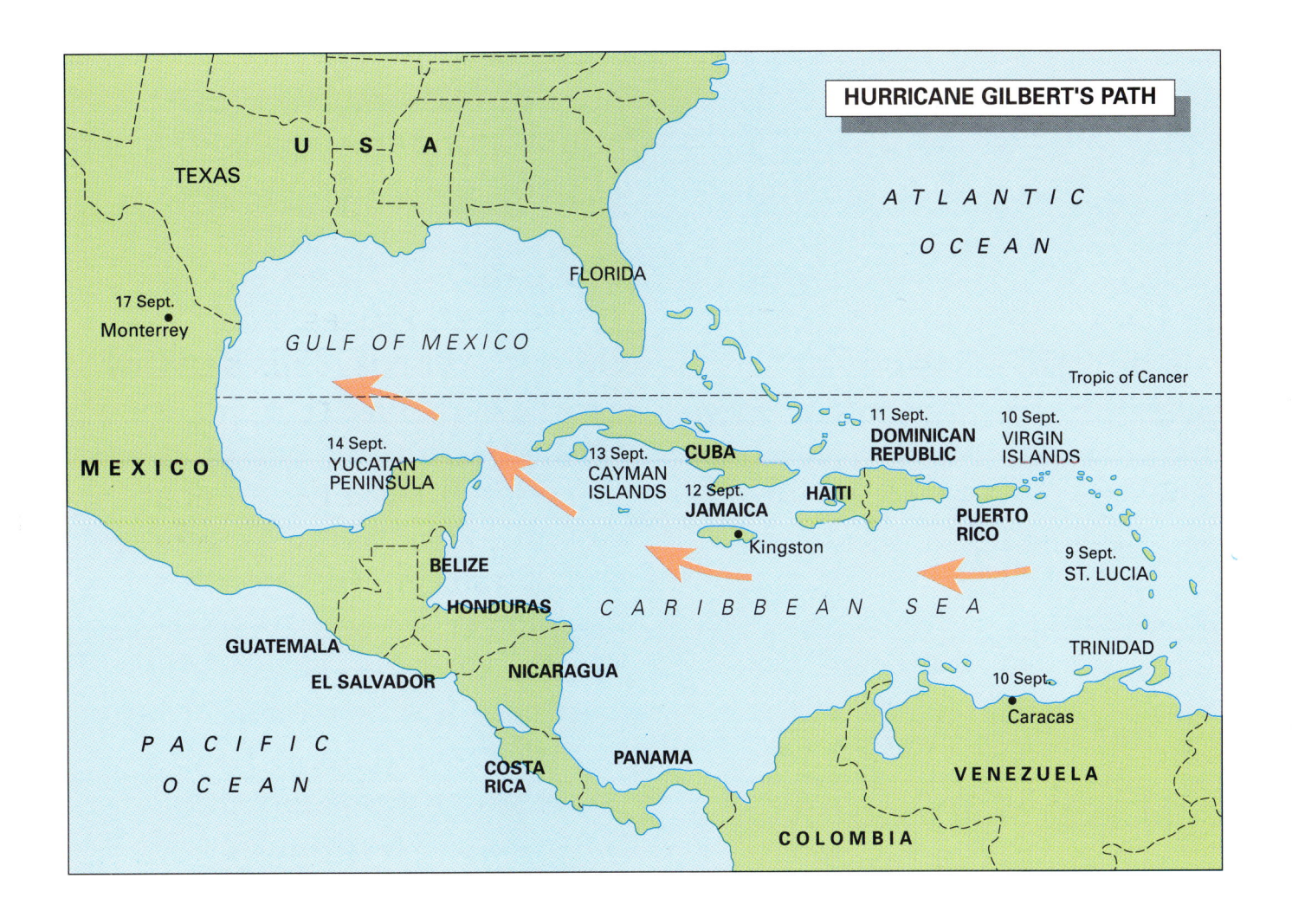

The beginning
Friday 9 September – Gilbert passed over the island of St Lucia (see map). The speed of the wind was only 50 kilometres per hour. But the wind destroyed the crops of bananas growing on the island.

Saturday 10 September – Gilbert moved across the sea. The hurricane was getting stronger. The southern edge of the hurricane touched Caracas in Venezuela. At least five people died. The northern edge of the hurricane moved across the Virgin Islands and Puerto Rico. It killed many farm animals and destroyed farmers' crops.

Sunday 11 September – The speed of the wind had now reached 120 kilometres per hour. At sea, ten fishermen drowned after their boat turned over. Gilbert moved across the Dominican Republic and Haiti. At least 27 people died. The hurricane destroyed many thousands of homes.

The middle
Monday 12 September – Gilbert changed its direction a little. This meant that it passed right over the island of Jamaica. The wind speed reached 225 kilometres per hour. The wind blew the roofs off almost all the houses on the island. Some of the local people described what happened, and the effect of the hurricane:

> 'I was trembling and crying for mercy and help. The tin roof could have cut off most of our heads.'

> 'Everything in sight is crushed. My wardrobe, my bed, it's all gone. We can't get anything out.'

Tuesday 13th September – Gilbert moved across the Cayman Islands. At sea, 80 fishermen died when five fishing boats sank.
Wednesday 14 September – Gilbert reached the mainland of Central America. It hit the Yucatan Peninsula in southern Mexico. The wind speed was almost 300 kilometres per hour. The hurricane damaged tourist hotels along the coast. At least 21 people died.

The end
Thursday 15 September – Gilbert seemed to be moving towards Texas in the USA.
Friday 16 September – Gilbert started to lose power, although the wind speed was still 200 kilometres per hour. People on the south coast of the USA got ready for the arrival of the hurricane. Many people moved away from the coast to safer places inland.
Saturday 17 September – Gilbert suddenly changed its direction. Instead of passing over the coast of the USA it moved inland over Mexico. The people of Mexico were not ready for the hurricane. There was flooding and the wind destroyed many buildings. In the centre of the city of Monterrey floods swept away four buses. Many people on the buses drowned. Six policemen also drowned as they tried to rescue the people on the buses.
Sunday 18 September – Gilbert finally died away. The people of the USA moved back to their houses.

▷ Hurricane Gilbert moved across the Yucatan Peninsula in Mexico. The hurricane destroyed many tourist hotels on the coast (below).

Protecting lives and land

Staying safe

In a thunderstorm the crash of thunder often makes people afraid. But it is the lightning that is dangerous. Even then, thunderstorms are a lot less dangerous than hurricanes or tornadoes.

You can stay safe in a thunderstorm by following a few simple rules. The most important thing is to stay away from trees. You should not shelter under a tree to get out of the rain. This is because the wind might blow the tree over.

There is another reason why you should not shelter under a tree in a thunderstorm. When lightning moves from the bottom of a thundercloud to the ground (see page 15), it hits the tallest object on the ground. So lightning could easily hit a tall tree.

A tall building has to be protected from lightning. This is done with a lightning conductor. On the highest part of the roof there is a metal rod. A metal strip runs from the metal rod, down the side of the building to the ground. If lightning hits the rod, the electricity in the lightning is taken safely down the metal strip to the ground.

Forecasting weather

It is often useful for people to know what the weather will be like. This is called weather forecasting. It is especially useful for sailors and fishermen. This is because

▷ This lightning conductor is used by scientists to study lightning. The rocket (inset) carries a metal wire into the thunder clouds to attract the lightning.

▷ Weather forecasts are useful for fishermen.

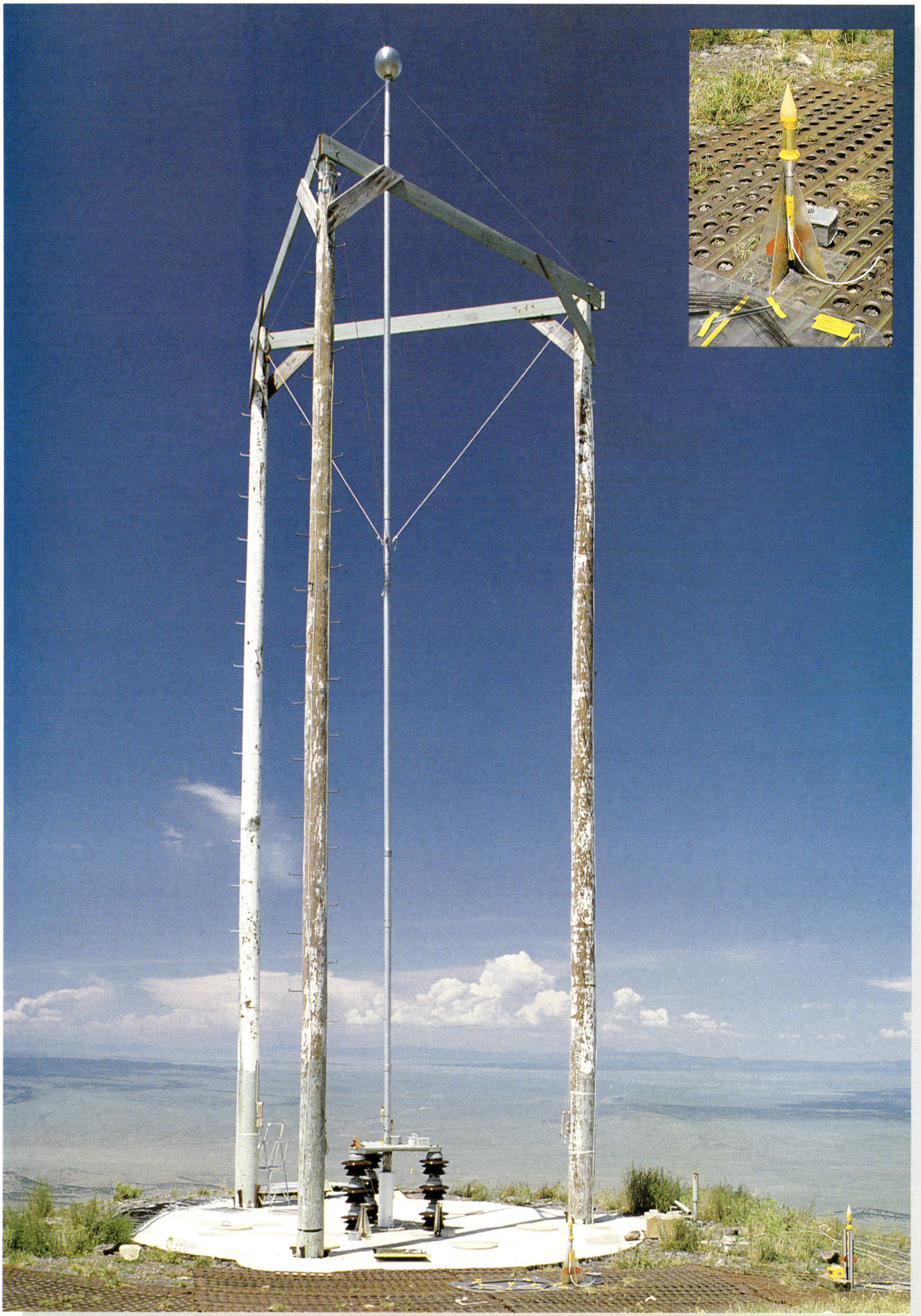

storms at sea can be very dangerous for small boats.

Special weather forecasts are broadcast on the radio for sailors and fishermen. They are called shipping forecasts. The shipping forecast tells people about the speed and direction of the wind. It gives information about visibiity. This is how far you can see. If there is fog or mist then visibility is poor. If it is a clear day then visibility is good. The shipping forecast also tells sailors how rough or calm the sea is.

Storm warnings

If a storm is on the way a gale warning is broadcast on the radio. A gale is a bad storm with strong winds and heavy rain. Warning about a gale will give sailors time to get ready, or return to port.

In places where there are hurricanes, hurricane warnings are broadcast on the radio and television. There are usually two warnings. There is one warning several days before the hurricane is expected. This is called a hurricane alert. It tells people to be ready if the hurricane moves in their direction.

Hurricane parties

After a hurricane warning the worst thing you can do is hold a hurricane party. In the 1950s people in the USA held these parties. Instead of moving away from the coast, people asked their friends to their houses to watch the wind and waves. Sometimes these party-goers were lucky. Other times they were not, and they were hurt or even killed by the hurricane.

When the weather forecasters know where the hurricane is likely to go, they broadcast the second warning.

If a hurricane is heading in your direction, the best thing to do is to get out of the way. People often move out of towns and cities if a hurricane is expected.

Protecting the coastline

The most dangerous place to be during any kind of storm is the coast. Hurricanes are the only storms that suck the sea up and carry it on to land (see page 20). But gales can sweep large waves on to beaches. The waves can flood towns on the coast.

There are many ways to protect the coast. People build walls called sea walls. Sea walls help to stop seawater from flooding the land along the coast.

△ In the Netherlands, concrete sea walls protect land along the coast.

▽ People in Vietnam build a mud wall to protect their land against flooding.

Modern sea defences

Some sea walls are solid concrete walls. But scientists have discovered that walls with holes along the front make better sea walls. The waves rush into the holes and the water swirls around. This takes energy out of the waves. This means that the sea wall will last longer than a solid wall, and it probably will not need as much repair.

Plastic seaweed
There are many oil rigs in the North Sea, off the coast of Scotland. Many of the oil rigs stand in water over 50 metres deep. If a storm is coming they cannot be moved out of the way. But rough seas and large waves could knock an oil rig over. So scientists have come up with an unusual answer. They have put large plastic mats on the sea bed underneath the oil rigs. These plastic mats have long ribbons of plastic attached to them. There are thousands of these ribbons. Together they take energy out of the waves

around the oil rig. They help to make the sea calmer.

Sheltering from tornadoes

It is difficult to forecast tornadoes. This is because a tornado comes out of a storm cloud without warning and usually lasts for less than an hour. In Tornado Alley in America (see page 10) most houses have tornado shelters.

These are underground shelters. They provide safe shelter if a tornado is coming towards the house.

▽ A tornado shelter in Texas, USA

▽ Plastic seaweed (inset) can protect oil rigs from storm waves.

Weather forecasting

Weather forecasts warn us about storms and hurricanes. But 100 years ago, there was not much weather forecasting. People took information from one place to another on horseback. The telephone and telegraph were only used in a few big cities. So there was no way of warning people of storms, either on land or at sea.

During the last 100 years, people have developed many new ways of sending information. This means that weather warnings can be sent to ships, and from one country to another.

Reports from sea
Radios were first used on ships in the early 1900s. With a radio, the crew in one ship could talk to people in another ship, or on shore. So ships could warn each other about bad weather. They could also send information about the weather in the middle of the ocean back to people on the land. For the first time, people could find out about hurricanes before they hit the land.

▷ Thick clouds cover the sky. Weather ships and buoys (inset) send information about the weather at sea back to forecasters on land.

35

Reports from the air
The first weather reports from the air came from weather balloons. Scientists sent measuring equipment high up in these balloons. But aeroplanes gave scientists much more information about the weather. By the mid-1940s, planes from the US Navy flew regularly to give warning of hurricanes. Planes can see much more of the ocean than a weather ship.

▽ A weather balloon (inset) and a weather plane

Satellites

Scientists can now watch the weather from space by using weather satellites. The satellites fly around the earth. They send back pictures which show the patterns of the cloud above earth. Hurricanes are easy to spot because they have a spiral pattern (see page 8). Satellites can also send other information, such as how hot or cold the air around the earth is.

▽ The European Remote-sensing Satellite (ERS-1) is a weather satellite. The map of the earth (inset) was made by a weather satellite. It shows the temperature of the air around earth.

Hailstones

Layers of ice build up to make hailstones. You can see these layers in the picture below. It shows a large hailstone that has been cut in half.

Weather maps

Weather forecasters get information about the weather from weather ships, satellites, and from weather stations around the country. All this information helps to make a weather map. A weather map shows areas of high and low pressure. This tells people about the direction and the speed of the wind. Weather maps also show areas of warm and cool air. An area of air that is warmer or cooler than the air around it is called a front. A warm front usually means that temperatures will rise. A cold front usually means that temperatures will fall.

Hurricane off course

Today, satellites send down pictures to show us which way storms and hurricanes are moving. But in the past forecasting the weather was not so easy.

In September 1938, forecasters gave a warning about a hurricane moving towards the coast of Florida, USA. People moved inland to safety. But nothing happened. The hurricane moved a different way. Nobody noticed what was really happening until it was too late. The hurricane moved north and hit the coast near New York. The people there had no warning and more than 600 people died.

▽ Damage caused by the hurricane of 1938 near New York, USA.

Climate and weather

The weather is changing all the time. The weather can be cold, warm, dry or wet. The pattern of weather in a place is called climate. We say that tropical countries have a hot climate. Arctic countries have a cold climate.

The climate of the earth has changed many times. About 100,000 years ago the climate of the earth was much colder. Most of Europe and North America was covered in ice and snow all year round. This is called an Ice Age. Today, the climate of the earth is getting warmer. Scientists think that this is happening because people have harmed the earth's atmosphere.

Warming the atmosphere
Scientists think that air pollution is harming the atmosphere around the earth. Air pollution comes from smoke from factories and fumes from cars. The most dangerous kind of pollution is invisible. It is a gas called carbon dioxide. Carbon

dioxide is made when we burn wood, coal, gas or oil. The carbon dioxide in the atmosphere traps heat from the sun. The heat cannot escape. This keeps the earth warm. But if there is too much carbon dioxide in the atmosphere, the earth could get too warm.

◁ Smoke and fumes from factories and cars pollute the atmosphere.

If the atmosphere around the earth gets too warm the ice caps at the North and South poles could start to melt. This would meant that there was more water in the seas and oceans. The level of the seas and oceans would rise. The extra water would flood many coasts and islands. The warmer atmosphere would change the climate on earth. No one knows exactly what would happen.

Controlling the weather
People have always dreamed about being able to control the weather. It would be useful to make it rain when crops need water, or to stop the rain when the crops need sunshine. Scientists have discovered a way to make it rain or snow. It is called cloud seeding. Planes scatter chemicals into large clouds. Rain or snow forms around the chemicals and falls to the ground.

▽ A warmer atmosphere would make the ice caps melt.

◁ In the early 1900s, farmers fired gunpowder into the clouds. This was to try to make it rain before hailstones formed in the clouds. If hailstones fell, they would damage the crops.

Scientists have used cloud seeding to try to stop hurricanes. But cloud seeding does not always work. Spreading chemicals in the atmosphere also causes more pollution. People will never be able to control all the world's weather. Instead we must do our best to protect people from storms and hurricanes, and help them to repair the damage afterwards.

Glossary

atmosphere – the layer of air that surrounds the earth.
climate – the weather conditions in a place over a long time.
concrete – a building material made from cement, sand and water.
equator – the imaginary line around the centre of the earth, half way between the North and the South poles.
front – the line where an area of warm air meets an area of cooler air, or cool air meets warmer air.
molecules – the tiny particles that make up all substances.
pollution – anything that makes the air or water around us dirty or dangerous to live in.
prevailing winds – winds that usually blow in a certain direction.
satellite an object that flies around earth and collects information to send back to earth.
tropical – from the tropics. The area called the tropics lies between the Tropic of Capricorn and the Tropic of Cancer (see map on page 5).
weather forecast – a report that says what the weather will be like for the next few days.